Earth
Plant
and
Compost

Earth
Plant
and
Compost

Edited and annotated by
William F. Brinton
Woods End Research Laboratory

Biodynamic Farming and Gardening Association

Copyright © 2002 by The BIO-Dynamic Farming and Gardening Association, Inc.

All rights reserved. No part of this publication may be reproduced, stored in a retrieval system, or transmitted in any form or by any means, electronic, mechanical, photocopying, recording, or otherwise, without the prior written permission of the publisher. For information about permission to reproduce selections from this book, contact the Biodynamic Farming and Gardening Association, Inc., PO Box 29135, San Francisco, CA 94129-0135.

Printed in the United States of America by Thomson-Shore, Inc.

ISBN: 0-938250-107-8

Portions of this book appeared originally as *Kompostieren: Anleitung für eine sinnvolle Verwertung von organischen Abfällen* by Alex Pfirter, Arnold von Hirschheydt, Pierre Ott, and Hardy Vogtmann. Permission for use granted.

Editorial staff:
William F. Brinton
H. Clark Gregory
Jonathan W.Q. Collinson
Mary Lynn FitzSimons
Heinz Grotzke

Eric Evans
Mary L. Droffner
Richard B. Brinton
Charles Beedy
Elisheva Kaufman

Illustrations by Margaret B. Collinson

Cover photograph – *On-Farm Composting: Bio-Abfell Site;* copyright © 2002 by William F. Brinton

Table of Contents

Preface vii

General Principles of Composting 1

Source Ingredients for Composting 14

Composting Methods 24

Sheet or Surface Composting 34

Evaluating Compost Quality 36

Compost Uses 45

Compost Mulching 52

Composting for Kitchens and Family Gardens 55

Special Composts 58

Glossary of Technical Terms for Compost and Soil 63

Compost and Related Bibliography 68

Useful Addresses – Technical Assistance 72

Preface

The roots of this booklet lie in Switzerland more than twenty-five years ago when as a graduate student I went to Europe to examine biologically oriented farm research opportunities. The US EPA had just been formed and there were no official documents at all available in America concerning composting. At that time, Hardy Vogtmann, now president of the German Nature Conservation Agency, was co-directing the small and struggling Oberwil Research Institute for Biological Husbandry (FIbL), today a large, federally financed organic research and certification center. Early work in Oberwil included compost field trials on farms and measuring compost effects on quality of vegetable crops – to name just some of the exciting research opportunities that lay waiting for a young student.

Enormous strides have been made since this time even while the basic tenets of composting have changed very little. Today, Europe and America together manufacture more than ten million tons of compost per year. Compost products, now available in virtually all garden centers, are used for all purposes from peat substitutes to garden fertilizers – indeed, even compete with fertilizer. And composting has become much more than the organic farming and gardening issue it once was; in fact, more compost is used today by conventional growers than organic. In this way, compost itself has helped dissolve the sharp boundaries that existed between organic and non-organic.

No topic is ever complete, and compost work continues to evolve significantly. While holding to basic tenets we must all notice the changes of practice and improvements in understanding that are taking place. If the nineteen-eighties and nineties were a time for quantitative developments in composting, the first decades of the twenty-first century will be the time to discover the fine qualities of compost. Through composting, and the systematic re-discovery of microbial biology, we may just be on a journey to acknowledge and appreciate our healthy roots in fertile, humusy soil.

William Brinton, Maine 2002

General Principles of Composting

Why make compost?
During the last few decades, soil fertilization practices have been largely oriented towards applying soluble mineral fertilizers as the sole basis of plant nutrition. Natural processes intrinsic to the soil-plant eco-system were hardly considered. Not only this, but if and when natural fertilizers such as farmyard manure and organic residues were incorporated, their contributions to the soil nutrient status were ignored by official sources when calculating crop fertilizer requirements. The organic materials were simply considered undesirable 'wastes', to be applied to soil as a means of 'disposal'. Potential environmental problems resulting from this practice were disregarded. So much for the past.

Several factors have caused a gradual but large shift in thinking about the soil-mineral theory. The increasing degradation of soils, estimated at a present loss of billions of dollars per year in America, has required greater use of fertilizers. However, the rising cost of fertilizers has led, among other factors, to stagnation of development. Moreover, crop yield returns have diminished as fertilizer rates have increased. Soil 'tiredness' is a new phenomena resulting partly from microbial impoverishment. Finally, there has been a growing awareness of environmental responsibility among large segments of the population, placing new pressures on the growers.

For these and other reasons, manures and organic residues are once more in the limelight, only now in a positive manner. The flip-side of the coin is that everyone is aware of the solid-waste crisis. Not for nothing, the processing and application of organic wastes, such as farmyard manure, bark, vegetable residues and even municipal garbage and sludges are currently on the forefront of agricultural development.

This change in attitudes favoring recycling of organic materials has resulted in renewed appreciation of wastes as 'resources'. Furthermore, the

Fig. 1.0 – Composting is an integral part of the balanced agricultural lanscape.

direct connection to the maintenance of soil fertility and productivity is being noticed. Unfortunately, it has also been noticed that soils are excellent disposal routes. To achieve true viability of the emerging composting industry means that organic wastes must be properly processed and correctly used. Like anything, this is more easily said than done. Thus, composting is at the center of a whole series of issues that challenges tradition as well as prevailing economic forces.

To successfully implement composting, several objectives must be achieved, as follows (Vogtmann and Besson 1978):
- Control and reduction of unpleasant odors associated with raw wastes;
- Improvement of waste hygienic; reduction of potential pathogens;
- Reduction of the germination capacity of weeds;
- Preservation and improvement of the nutrient status;
- Increase of the biological activity of soils;
- Positive influence on plant quality;
- Minimum loss of nutrients during application;
- Acceptable working conditions;
- Minimal energy input requirements;
- Costs advantages compared to other disposal methods.

Each year, an enormous quantity of organic matter, renewed by photosynthesis, undergoes natural microbial transformation and becomes partially preserved in the soil in the form of humus. This 'humification' process normally occurs very slowly in the natural environment on and below the soil surface, at ambient temperatures.

The decay of plant tissues discarded on the earth's surface is a natural process which occurs without any human intervention. Growers exploit this natural principle when they surface compost. This practice, also called sheet composting, refers to the deliberate spreading of various raw residues on the soil surface and is used in numerous agricultural situations. In fact, it is the basic means through which most organic materials, including manures, have been traditionally utilized in farming and gardening.

Composting differs substantially from sheet composting and fulfills a special role. For example, when organic residues like manure and harvest residues accumulate on a continual basis, the immediate application to soils is difficult, if not impossible. Under sub-optimal conditions of wet soils or in winter on frozen ground, surface spreading of manures and other wastes is in fact undesirable if not illegal. In these circumstances, composting offers a practical solution to storage and improves the natural value of such materials before soil spreading.

The composting process

Proper composting begins with the selection and mixing of initial raw materials. Basically, the composting process depends upon establishing a balanced relationship of organic matter, nutrients and air-space. Given these conditions, the proper microbial processes will take place with little needed intervention.

Composting requires a more deliberate approach than is needed when simply land-applying wastes, but it would be misleading to say that composting must be tightly controlled. Rather, the correct 'environment' must be created in a holistic sense. By creating optimal conditions, an environment favorable to the development of bacteria and other necessary organisms will likely lead to successful composting.

While it is generally true that compost will work once the basic conditions are established, there are of course several aspects which need to be properly understood to achieve success consistently. The balance of materials in terms of the ratio of carbon to nitrogen (C:N) in the source ingredients is one of the most frequently cited parameters determining the outcome of the composting process. In actual fact, C:N is greatly overrated and can actually vary widely without adversely affecting the process. The type of organic materials making up the mix is easily as important. Less frequently mentioned but of great significance is the porosity of the mix. This determines how well the material can be aerated – meaning how well air may diffuse into the mass. The amount of aeration (oxygen supply) should be carefully anticipated. Finally, the moisture content of the mass is of great importance, and this is relatively easily controlled at the outset.

Porosity, aeration, and moisture are directly related with each other. Excessive moisture or poor porosity reduces air diffusion and can seriously hurt the process, causing putrefaction, which among other things, leads to bad odors.

With regard to C:N ratios, the ideal C:N balance of a compost based on farmyard and/or market garden wastes varies generally between 25 and 35, meaning about 30 parts of carbon to each part of nitrogen. C:N ratio is a very impractical guide because it can't be readily measured. Even if it is tested, the actual availability of the carbon can never be fully known.

One thing is certain about C:N ratios. If the ratio in a compost is too low, say less than 12:1 at the start, then the compost is likely to smell poorly, or give off excessive ammonia. Should the ratio be much higher, the excess carbon material will require a greater number of microbial generations for its degradation, meaning the compost will take much longer to be done but it will still get there. This, of course, is not necessarily a problem when composts are made and applied the following growing season, as in the case of farmers and gardeners. The modern emphasis on industrial composting has caused the shift in favor of controlling the C:N ratio more closely to increase the speed of decomposition and reduce processing costs.

Composting would not be composting without the noticeable rise in temperature, but the amount of heating is relative, and compost can be successfully made at any temperature above freezing. Plant residues and the excreta of grazing animals, which are surface-composted, i.e., decomposed on the soil surface, will degrade without any apparent temperature increase. Actually, all degradation produces energy, only we can't measure it under ordinary conditions. When, however, the organic residues are piled-up, then the energy produced by microbial degradation will not dissipate as fast as it is produced. This heat accumulates inside the pile, resulting in the rise in temperature considered to be a hallmark of composting. It is hard to imagine composting without a rise in temperature. The quantity and duration of heating affects the entire process. For one, the microbial populations shift to bacterial genera ('thermophiles') which are capable of growing under high-heat conditions. The higher temperature produces 'positive feedback', increasing the microbial reproduction rate to as frequent as a generation every twenty minutes! Under these conditions of high respiration rates, composting is quite different from normal biological degradation processes as they might occur in the soil.

The ability of a compost to generate heat and sustain high temperature depends among other things on the following factors:
- The composition of physical and biological components;
- The availability to micro-organisms of carbon and nutrients in the mixture;
- The level of moisture in the source ingredients;
- The structure of the material (particle size and texture);
- The rate of aeration of the compost-heap or windrow;
- The size of the compost heap or windrow.

Figure 1.1 provides a simplified graphical representation of the composting process. The various components of the process are shown in terms of inputs and outputs in relation to intrinsic factors. In some cases, some of the important factors are both intrinsic and extrinsic, meaning some are added on top of what is already there. For example, microbes are present in all materials, but often more are added by applying soil, ripe compost, or certain kinds of starters. The same goes for moisture, which is both present in the initial wastes, added during the process, and is a significant output in the form of moisture vapor.

The nutritive elements such as proteins, amino-acids, lipids and carbohydrates are present in the wastes and easily assimilated and broken down by the micro-organisms. These ingredients supply 'energy' and after undergoing various degradation steps, are recovered on the other side of the equation chiefly as carbon dioxide (CO_2) and water (H_2O). More

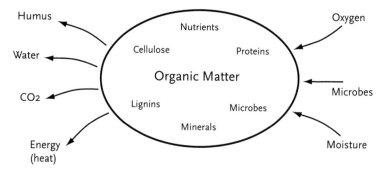

Fig. 1.1 – The composting process

difficult to degrade cellulose, lignin, and the non-decomposable inorganic (ash) fractions contribute to the formation of humus.

It may be hard to imagine, but composting is a kind of controlled burning that does not go to completion, otherwise only ash would remain. Instead, some of the organic matter remains or is converted to humus – a dark brown, amorphous substance that is the true basis of lasting soil fertility and is very resistant to further decay.

Micro-organisms play a key role both in the degradation and the formation of humus through synthesis of resistant compounds and the sloughing of their own tissues. The rate of microbial reproduction depends largely upon temporary conditions, including temperature as well as the supply of nutrients and moisture. The complex metabolism of the micro-organisms releases heat, carbon dioxide (CO_2), water (H_2O), oxygen (O_2), hydrogen (H_2) and other intermediary compounds until the process is complete.

Traditionally, the composting process is divided into several typical phases or stages encompassing heat-up, rapid-decay, and final stabilization. One may speak of two to four distinct phases, depending on the point of view. The three phases most customarily cited in literature are:
- A degradation phase (high heat with rapid break down);
- A transformation phase (heat subsidence);
- A maturation or curing phase (product stabilization).

Unfortunately, these terms are very arbitrary. Whether there is a biological basis for them is uncertain and will depend on the viewpoint of the observer. For example, a microbiologist may notice the large differences in populations of microorganisms between early and later compost stages. Thus, to a microbiologist, there are at least two distinct phases of composting. A chemist might be inclined to consider the process purely from the point of view of energy release and nutrient transformation. The latter has tended to dominate the agricultural point of view until recently.

Heat build-up is certainly one of the most remarkable features of the composting process. During the degradation phase, the temperature in the center of the heap may reach, and even surpass, 165 degrees F (see Figure 1.2). This is of great importance for the destruction of pathogens and especially of weed seeds. Depending upon the composition of the material being composted, maximum temperature is reached between two and fifteen days after the composting process begins.

On the surface of the pile, however, the temperature seldom exceeds the ambient temperature. It is common practice, therefore, to cover it with a layer of straw, wood-chips or mature compost in order to expose all of the fresh materials to the higher temperatures prevailing inside the piles. This is not necessary with large windrow operations where the piles are being regularly turned. In this case each turning of the pile brings the previous surface materials into the warmer interior, thereby causing the desired weed and pathogen destruction.

Pathogen destruction should not be thought of in the passive sense,

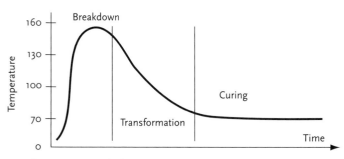

Fig. 1.2 – Temperature cycle during composting

as simply destruction by high temperature. Rather, a combination of factors, including microbial competition, antagonism and antibiotic effects, in addition to heat, act to reduce pathogenic organisms that may be present in the compost ingredients. The best guarantee of a safe compost is one which has gone through the proper stages of heating and cooling followed by curing under ideal conditions.

During the degradation phase, mesophilic and thermophilic bacteria play a prominent role in the major transformations. Their activity is matched to a much lesser degree by actinomycetes (filament-shaped bacteria), and even less so by fungi. However, the distinction of bacteria into two temperature groups is somewhat arbitrary, as many microbes are able to adapt to altered temperatures.

During the so-called transformation phase, temperature starts to drop and other bacteria and/or fungi may complete the decomposition. During the maturation or curing phase the bacterial activity slows considerably, and under suitable conditions, the pile may become gradually inhabited by earthworms, springtails and mites. It is unlikely that this macro-fauna will exist at all following a very hot, intensive composting period, particularly if the compost piles are indoors or on a compacted floor.

Contrary to popular opinion, there is no specialized group of microorganisms that carry out composting, but rather a great variety of organisms fulfill the task dynamically. Bacteria that grow at low temperatures (mesophilic), can readily adapt to thermophilic (over 120 degrees F) conditions. Furthermore, many thermophilic fungi and actinomycetes

Table 1.0 – Microorganisms observed in the composting process

Organism	Typical quantity per gram of compost
Bacteria	$10^4 - 10^9$
Actinomycetes	$10^2 - 10^6$
Fungi	$10^2 - 10^6$
Algae	up to 10^4
Protozoae	up to 10^6

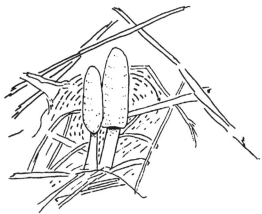

Fig. 1.3 – Development of mushroom fungus is often observed on a manure compost that contains some straw or woody residue

may persist in very hot, dry composts. The ashen, dusty look of such composts, often referred to as 'burning', actually indicates thermophilic actinomycetes. These strange bacteria tend to grow well where conditions are inimical to other, more aggressive bacteria, which prefer more moist conditions.

The danger of over-simplifying the microbial picture of composting is illustrated in the popular concept of aerobic versus anaerobic organisms. Compost is considered to be 'aerobic'. However, the most prevalent bacteria in composts are 'facultative' organisms which will function aerobically until oxygen gets too low, at which point they continue their metabolism anaerobically. This ability has probably enabled these organisms to better survive changing environments, making them well suited to composting. In any event, the microbial composition is far more complex and less well understood than one might be led to believe, but one thing is certain: it works!

The composition of a mature compost reflects the raw materials from which it is derived. To illustrate this, the Table 1.1 on page 10 compares a municipal solid waste compost and a compost made from farmyard manure.

Compost traits will vary widely, depending on the source ingredients

Table 1.1 – Average percent composition of two mature composts

Variable	Municipal solid waste	Farm yard manure compost
Organic matter	33	60
Physical contaminants	17	< 0.5
Nitrogen (N)	0.8	2.8
Phosphorus (as P_2O_3)	0.9	2.2
Potassium (as K_2O)	0.6	2.6
Calcium (as CaO)	7.3	3.1
Ash (mineral fraction)	67	40
Salt (conductivity)	1.5	8.0

and conditions of composting. For example, nitrogen and organic matter content may depend entirely on processing conditions and age which are independent of the initial starting materials. Researchers have demonstrated a surprising similarity in the composition of humus from compost materials of widely varying origin (Brinton 1979).

Macro-organisms which inhabit composts

A variety of macro-organisms can also be found in composts (although they are often overlooked) and are mostly found in the later stages after heat has subsided. The presence of these organisms is helpful but not essential to composting. Figure 1.4 illustrates some of the observed macrofauna that will typically inhabit composts (Bockemühl 1988).

Advantages of compost

Agricultural field trials comparing the use of compost versus fresh manures and chemical fertilizers demonstrate significant advantages for compost based programs such as the enhancement of soil humus, reduction of nitrate leaching, and improvement of the supply of available phosphorus (Eichenberger 1979; Maynard 1993; Vogtmann 1994).

Experiments carried out in Europe have shown consistently above

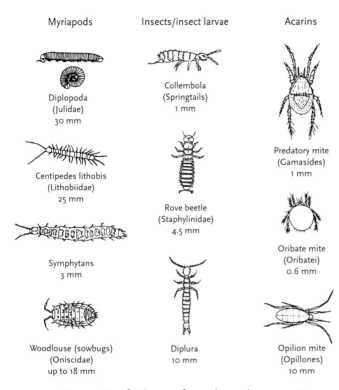

Fig. 1.4 – Beneficial macro-fauna observed in composts

average yields in a number of crops when a compost-based instead of mineral-based fertilizers are used over time. In other cases, increased nutritional quality and improved crop shelf life result from use of composts (Linder 1984; Pettersson 1984; Schuphan 1972). Alfalfa longevity has been shown to significantly improve with compost, while being reduced when raw manures or some mineral fertilizers are applied. Long term trials conducted in Sweden using various soil amendments over thirty-seven years, have demonstrated the superiority of composts over raw manures and mineral fertilizer blends with regard to soil condition and crop protein quality (Pettersson 1984).

One of the most often cited disadvantages of composting is the loss of nitrogen during the high-heat phase. While it is possible to keep such losses to a minimum through proper composting, the trait must be seen in context. Raw manures as well as chemical fertilizers have been shown to lead to large nitrogen (and other nutrient) losses by leaching, de-nitrification, and volatization. These losses are generally great enough as to wipe out any apparent differences between the groups. Not only this, but the undesired or harmful environmental effects of uncomposted raw wastes or chemical amendments must be taken into account. An example of this is the release of noxious ammonia and leaching of nitrates into ground water. Numerous field studies have shown less nitrate leached from soil-applied compost than from manures or chemical fertilizers, even at equivalent nitrogen rates.

Nitrogen loss from composting is correlated primarily with the extent of organic matter breakdown. Methods which enhance rapid and high-temperature degradation will invariably lead to increased losses (USDA 1993; Jobin 1992). By controlling composting so that moisture or heat extremes are avoided, the conservation of nitrogen and organic matter is accomplished. An increasing amount of research is showing how composting methods can be adapted to reduce nitrogen loss and eliminate potassium leaching.

When evaluating the efficacy of a compost, a wide range of factors need to be considered. While it is customary to look at mineral elements (N, P, K), this focus overshadows other important advantages, such as:
- Improvement of soil structure;
- Reduction of runoff, erosion control and lessened leaching of nutrients;
- Increased diversity of insect and microbial population;
- Reduction and suppression of soil transmitted diseases.

Investigations carried out by Schaerffenberg (1968) have shown that potatoes grown on compost may be less infested by the Colorado potato beetle. Results from investigations at the Biological Agricultural Research Institute at Oberwil (Switzerland) have demonstrated that vegetables grown on compost contained significantly lower amounts of nitrates, increased levels of vitamin C, and higher dry matter contents. These trial results have been confirmed by Schuphan (1976) and Lairon (1984).

Interesting compost trial results obtained by Cook (1976) proved that avocados grown with organic fertilizers and compost were less attacked by phytophthora (blight) than those fertilized with only mineral-based fertilizers. Chaboussou (1978) and Tränkner (1993) have clearly shown that crops fertilized with compost are generally less susceptible to disease and pests. In fact, Tränkner and Weltzien (1992) give clear indications of the disease suppressive nature of composts, particularly extracts or compost 'teas' prepared from matured composts (see page 49 in the chapter entitled "Compost Uses").

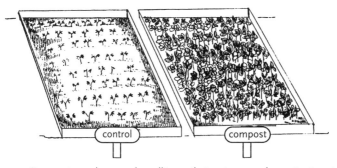

Fig. 1.5 – Compost supplemented seedling soil gives improved germination, improved growth response, and natural disease control.

Source Ingredients for Composting

The source ingredients which comprise the raw 'feedstock' for composting are typically wastes or by-products of a completely organic or natural composition. The expressions 'raw materials', 'wastes', 'resources', and 'residues' are used synonymously in composting. (To some extent, entirely new jargon has come to dominate the discussion of composting, encompassing ecological, environmental, and even activist orientations to the subject.)

Traditionally, the key ingredients to compost are those wastes resulting from normal activities in farming, gardening, or food processing. However, the boundaries of acceptable ingredients are expanding as composting priorities now include industrial residues, municipal sludges (now referred to as 'biosolids'), contaminated soils, and even chemical compounds such as herbicides and pesticides.

A wide variety of traits are important in selecting compost ingredients. The process begins with assessing the degradability, potential contamination, structural integrity and nutrient composition of potential ingredients. Successful composting depends on the right proportions of physical, chemical, and biological traits contributed by the various materials. There is no single formula to guarantee a good compost. As in good gardening or farming, some discretion is required, and one should avoid overly-simplified recipe schemes which are popular in the literature. The following points may be helpful in making the proper selections.

The raw material is natural and is truly bio-degradable

Natural, 'organic', or bio-degradable refers to a substance formed by a prior living process or containing organic compounds which can be broken down readily in nature. Normally, compost ingredients consist of plant or animal matter in a non-adulterated state. Processed or man-made materials including resins, paper and cardboards, and to a certain extent the new natural plastics called biopolymers must be treated with caution. A rule of thumb for composting is to define bio-degradable materials as those needing not more than nine months to fully degrade, and the residues will tend to accumulate somewhere they don't belong.

Virtually all organic materials can be composted as is, but many must first be size-reduced by grinding or chopping and/or mixed with other materials to ensure a proper composting process. Truly natural ingredients generally do not possess any characteristics that would cause concern for a composting operation.

The raw material is natural but results from an industrial process
There are many compostable materials that are naturally derived but have been processed to a greater or lesser extent, or they are industrial products which appear to be essentially decomposable. These include municipal sludges, paper, and paper sludges from industry as well as unusual food-processing residues like alkali extracted potato skins, coffee hulls, and so on. Such ingredients may be composted, but require more care or caution than observed with strictly natural ingredients.

The raw material is of a mineral, non-organic nature
Unlike organic materials, mineral or non-organic materials contain little or no carbonaceous matter. Such mineral materials obviously can not be composted alone, but they can be advantageously mixed with organic materials to be composted, are usually not harmful, and are sometimes helpful. These ingredients include wood and waste ash, quarry dusts and mineral powders.

Mixing compost ingredients
The selection and mixing of organic materials for composting depends on the following factors:

STRUCTURAL COMPOSITION
Structural integrity of compost is the most important trait. The structure is determined by the physical nature of the ingredients and their behavior during the composting process. Some materials (sludge, vegetable

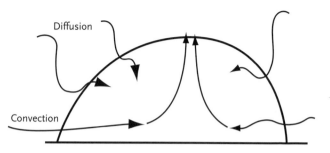

Fig. 2.0 – Air will both diffuse and be carried by convection into a compost pile containing sufficient coarse or bulky particles.

residues) either have little structure or rapidly lose structural integrity during composting. Other ingredients (wood chips, bark, straw) tend to retain their structure. Ideally, the mix will possess properties of both.

A well-structured pile allows air to penetrate both by natural diffusion and by convective forces. Both factors play an extremely important role in maintaining proper composting conditions. Any system or method that restricts one or the other of these forces must be compensated for by another. For example, putting a pile in a solid bin may severely restrict convection, even if porosity of materials is adequate. If that is the case, then one may have to provide aeration by adding air holes under the floor to compensate for it. Piling materials too high results in both compression and a significant reduction of diffusion and convection forces. Under these conditions, air may have to be blown into the improperly constructed pile.

NUTRIENT AND SALT CONTENT

There is no ideal level of nutrients, but the general principle is to get the C:N ratio right and assume everything else falls into place. Ultimately, the important ratios are C:N:P:S (carbon : nitrogen : phosphorus : sulfur). It is not necessarily true that a precise computation needs to be made, however. In order to obtain an optimum ratio it is desirable to mix unlike materials together, such that those rich in nitrogen are blended with others high in carbon. This is often referred to as mixing 'brown' with 'green' waste. Under certain circumstances, deficiencies in phosphorus and sulfur may have to be compensated for, but most organic materials derived from living processes have a fairly ideal ratio to begin with.

MOISTURE LEVEL AND WATER-HOLDING CAPACITY

The moisture content of raw materials is often either too high or too low to permit a suitable composting process. From a practical point of view, it is more economic to mix moist materials with dry ones, rather than to attempt to dry them out. It is possible, however, to spread wet materials thinly to a depth of six inches, and allow them to dry by natural wind or sun before heaping them into a compost pile. The addition of dry materials readily ensures a good mixture, as the moisture of the wet fraction is gradually absorbed by the drier.

UNDESIRABLE ELEMENTS

Some raw materials may contain significant quantities of heavy metals (Cd, Pb, Cr, etc.) or may be contaminated by undesired chemical compounds (PAH, PCB, pesticides, etc.)(Vogtmann et al 1989). When compost made from such ingredients is used for agricultural purposes, it is essential to understand the potential for permanent land contamination. Many contaminated residues such as herbicide-laced grass can be composted and completely broken down prior to subsequent use provided adequate time and proper conditions hold throughout. However, recent experience has shown that agri-chemicals such as herbicides designed for seasonal farm use may still be present in composts in active form even after several months of composting (Bezdicek 2001; Brinton and Evans 2002). For other contaminants, composting may be attempted but only in strictly controlled operations.

MINERAL ELEMENTS

When enriching a compost with inorganic materials such as wood ashes, the potential increase of nutritive elements such as potassium may be significant. Additionally, some ingredients may impart a lime-potential to compost. Compost made for turf application frequently has added iron.

Table 2.0 on the following pages lists a number of raw materials in relation to their above-mentioned characteristics.

Table 2.0 – Types and value of raw ingredients

Origin	C/N ratio & nutrients	Structure – Porosity	Moisture – as is	Degradabilit	Preparation/ Treatment	Limitation or dangers
Wood and lumber industry materials						
Bark	100–300; low: P, Ca; pH: low	very good	medium; good	very good	pre-grind	——
Paper sludge	100–110	medium to poor	very moist	medium	presscake	dioxins
Cotton sludge	20–40; N-rich; Low: P,K	poor	very moist	very good	pressed	——
Sawdust: Beech Fir Aged	–100 –230 <100	very good	<=50%; good	excellent medium poor	already ground	——
Cardboard	200–500	medium–poor	very low	very good	shred	boron, colors
Wood ash	n/a; K-Ca rich; high in heavy metals	poor	very low	none	none	metals, high pH
Municipal and residential residues						
MSW	30–120	medium–poor	very low	medium	grinding, moisture	metals, glass, &c
Biosolids (sewage sludge)	<20; high: P, N; low: K; metals	poor	high	very good	needs bulking	pathogens metals
Food scraps	<25; high: K, salt	very poor	high	very high	bulking	pathogens salt
Coffee grounds	30	medium	medium–high	low	——	slow breakdown

Table 2.0 continued – Types and value of raw ingredients

Origin	C/N ratio & nutrients	Structure – Porosity	Moisture – as is	Degradability	Preparation/ Treatment	Limitation or dangers
Garden landscape materials						
Wood chips	40–100;	good	too dry	low	grinding	coarseness
Garden waste	20–60	good	medium	medium	grinding	——
Green foliage	30–60	medium–good	good/dry	good	——	——
Leaves	50–70	good	variable	medium	——	matting
Grass clippings	12–25	poor	moist	high	bulking, pre-drying	odors
Reeds/ swamp matter	20–50	good	dry	medium	grinding	coarseness
Ditch scrapings	10–15	poor	moist	medium	drying	salts/lead
Agricultural residues						
Poultry manure (fresh, no-litter)	7–10	poor	moist	good	bulking	odor
Poultry manure (with litter)	13–30	medium	low–dry	medium	——	odor
Slurry (urine) liquid	2–3	poor	liquid	good	mix with dry matter	odor
Manure (cattle) liquid	8–13	poor	liquid	good	mix with dry matter	odor
Manure (pig)	5–7	poor	high	good	——	odor, moisture

Table 2.0 continued – Types and value of raw ingredients

Origin	C/N ratio & nutrients	Structure – Porosity	Moisture – as is	Degradabilit	Preparation/ Treatment	Limitation or dangers
Agricultural residues, continued						
Cattle manure	17–22	medium	medium	high	—	—
Manure with straw	25–30	good	good	medium	mix well	—
Horse manure	20–30	good	good	medium	moisten	—
Vegetable wastes	12–20	poor	moist	high	—	low pH, odor
Straw:						
oat/rye	60	good	dry	medium	chopping	—
wheat	75–100	good	dry	medium	chopping	—
barley	40–50	good	dry	high	chopping	—
Fruit pressing residues						
Grapes	poor in P, Ca	poor/ medium	medium	medium-low	lime addition	low pH, seeds
Fruits	poor in P, Ca	poor	medium	fair to good	lime addition	low pH
Others						
Peat (dark)	60–80	good	medium	very low	—	low pH
Peat (light)	60–80	good	medium	low	—	low pH
Slaughter wastes	15–18	poor	moist	high	—	odor
Mushroom compost	30–40	good	good	good/ medium	leach salts	salt fungicides
Rock powders	CA, K, Mg trace elements	poor	none	none	—	dust

Additives to the composting process

DEFINITIONS

Additives for composting include activators, microbial starters, and supplemental materials of either organic or mineral origin. Such additives are intended to aid or control the composting process and to alter and improve the quality of the finished product. For example, lime is often added to acid-type materials such as leavesand food scraps; gypsum can be added to manures to help avoid loss of ammonia.

ACTIVATORS AND STARTERS

When activators are utilized, they are generally used in very small quantities (<1 percent) to accelerate or improve the composting process. Very little is known about the efficacy of compost starters, and only two or three scientific studies have ever been conducted to evaluate the effects of commercial inocula, all inconclusive. Even if it were not so, it should be pointed out that organic residues usually contain very large populations of established organisms prior to composting. Therefore, adding organisms and expecting them to change the process means that they must overpower or suppress the existing populations, which is highly unlikely, except under very unusual conditions.

Three general types of activators or compost starters are known:
- Activators based on medicinal plants and biodynamic preparations;
- Activators based on bacteria and/or other micro-organism cultures;
- Enzyme based activators.

Although the efficacy of bacterial or enzymal compost starters for improving the composting process and the quality of the finished compost has never been fully demonstrated, there is some basis to believe that compost starters could be effective. This is seen with waste materials which are either fairly sterile, or resulting from an unusual process. For example, fresh food scraps are not heavily colonized by bacteria. Heat or steam processed food scraps or alkali treated bark or wood-pulp represent materials which have been partially if not fully sterilized. Under these circumstances, it may be well to add either a compost from a previous successful process or a compost activator. In a research trial Woods End scientists compared three types of composts to which bacterial

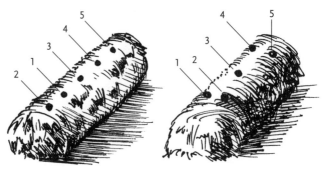

Fig 2.1 – The biodynamic herbal compost preparations are placed into holes in the compost pile at various locations.

starters had been added. The results showed that organisms added to manure composts had no effect, added to leaf and yard waste compost only a slight effect and when added to cafeteria food waste compost a very significant effect (Brinton and Droffner 1994).

The biodynamic compost preparations have been used by growers who employ the biodynamic method of farming as originated in 1924 by Rudolf Steiner in Silesia (now Poland). These specially prepared herbal/animal material compounds were intended to support the compost process in many ways, e.g., in controlling heating and avoiding undue losses of nitrogen. They are not viewed as microbial inocula and are not added to specifically increase microbial populations. Scientific trials have demonstrated significant effects, for example, on humification rate and plant quality after using the biodynamic compost (Goldstein 1982; Reganold 1993).

OTHER SUPPLEMENTAL MATERIALS

A variety of natural supplements are often used by gardeners during the composting process. The most often cited reason is to increase the content of minerals or purely to improve physical properties. The use rate of these supplements varies between about 1 percent up to 10 percent of the total mass. For example, gypsum can be added to manure at about 100 pounds per ton. On the other hand, seaweed powder should only be added at about the 1 percent rate, and wood ash at an even lesser rate of 0.2%.

The type and amount of additive should be determined in accordance to known requirements of the compost, or balancing an existing deficiency of the raw compost ingredients. In most cases, these supplements are kept near a compost pile and are sprinkled on each layer. The following table lists the various additives that are commonly used.

Table 2.1 – Possible additives to the compost process

Supplement	Description and Purpose of the Supplement
Activating compost	A compost used from a previous compost, which is not too aged and which is rich in organisms. Fairly large quantities are required (up to 10 percent of the mass) in order to achieve an effect.
Agricultural limestone	Used for correcting calcium deficiencies and moderately low pH values (of the compost mixture and /or of the soil to be enriched).
Blood meal	A high nitrogen organic fertilizer used in the absence of animal wastes (fresh or dried manure). It helps to produce a compost of excellent quality when manure is not available. May be restricted in some countries.
Bone meal	Mainly consisting of phosphorus and calcium, it is used to correct deficiencies of these two elements.
Clay soil or pure clay	Soil, montmorillonite, or bentonite is sometiimes used to enhance the formation of clay-humus compounds, as well as to improve the compost effect on sandy soil.
Horn meal	Its effect is similar to dried blood meal, but it acts more slowly, which is an advantage when being used in compost.
Gypsum	Added to control loss of ammonia by formation of ammonium sulfate and to improve soil texture after application.
Rock phosphate	These are natural or mined sediments. Finely milled phosphates are mostly water insoluble, and become available to plants over time through natural weathering by the action of soil acids and microorganisms.
Sand/coarse granite dust	Can be effectively used in small amounts to loosen texture and improve drainage. Part of its silicic acid, an important plant growing agent, is released by microbial activity.
Seaweed meal	This is a soil supplement made from freshly harvested seaweed. It contains large amounts of potassium and trace elements.
Rock meals	Act as a supplier of minerals (mainly trace elements) or clay. They can reduce unpleasant odor, favor the formation of stable humus, and improve drainage.

Composting Methods

Various backyard composting methods
THE FREE-FORM PILE
The traditional 'free-form' pile is a simple and effective method of composting. Basically, residues to be composted are mixed and layered into a single conical heap. Composting in open heaps has advantages as well as disadvantages:
- Monitoring of the composting process is easy, and turning is simplified;
- It is a simple matter to add fresh materials, either by spreading on the top of the pile or by extending it laterally to any dimension;
- Piles may appear messy and unsightly and attract animals.

When adding layers in open piles, it is typical that the materials do not degrade homogeneously. For this reason, it is recommended to stockpile course, low-nitrogen materials (leaves, brush) so that a supply will always be on hand to mix with fresh green wastes, which must be promptly incorporated into a compost.

There is no doubt that the loose pile method requires a larger area than the container or bin method; however, this should not be considered a

Fig 3.0 – 'Free-form' compost pile made of layers of household vegetable residues and yard wastes heaped in a cone-shaped pile without any support.

problem. When only 5–10 percent of a garden area is devoted to the preparation of compost, it should not be looked upon as excessive, particularly if we consider how much its productivity and its health depend upon good compost.

BARREL OR DRUM COMPOSTERS

A large variety of circular drums and closed bins are marketed in various designs made from wood, recycled plastic, aluminum, and steel. Compost barrels are recommended for composters who desire to rotate or mix the compost mass. Some drums are self-powered and turn at a preset rate.

As with any system, compost barrels have advantages and disadvantages. Unlike stationary bins, rotating barrels permit frequent mixing and improve the homogenization of the mass. While this may speed up the process, it does not guarantee it. On the other hand, drums do not permit soil-contact so no migration of earthworms into the mass takes place. Furthermore, if they are not powered, the rotation of such drums by hand requires considerable physical effort, so the size is often limited to about 5–12 cubic feet, which may seem small to an avid composter.

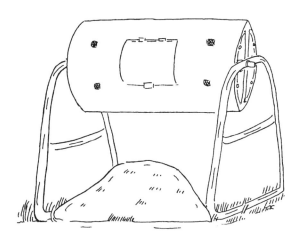

Fig 3.1. Compost barrel; a rotating hand-driven barrel made of recycled plastic that holds 30–80 gallons.

Fig 3.2. Wire latticed or solid-wall bins hold compost materials in a distinct location and form but are more difficult to turn or empty.

BIN COMPOSTERS

The most popular form of compost container is the so-called bin, which may be either an open-lattice system or a closed container. Like the compost barrel, a compost bin or trellis is particularly suitable for the backyard and small garden. Compost bins appear neat and tidy and require only a small operational space. Furthermore, it is relatively easy to place several bins side-by-side for an efficient operation.

Some kinds of open bins such as wire or wooden-lattice types may become overly dry from wind desiccation. To avoid this, they should be placed in the shade and watered frequently, as though they were are garden. The tops are often left open, but may need to be covered in very wet spells. The cover is raised when wastes are added, so it always protects

Fig. 3.3. Multiple-partition bin compost system used for a large garden operation or in a small institution like a school.

the last accumulated wastes on the pile's top. A gardener who wishes to produce different kinds of compost in small quantity will hardly be able to manage without this type of container.

Multiple partition connected bins made of wood are excellent for purposes of large gardens or institutional composting programs. In school composting programs that Woods End has developed in New England, a four-partition bin system is used for an entire school year of 250 pupils (MWMA 1994; see also page 56 in the chapter "Composting for Kitchens and Family Gardens").

Windrow composting

The word 'windrow' comes from farming where hay is loosely rolled into long rows on the ground for wind drying prior to baling. Windrow composting consists of making long piles of compost that resemble such rows.

Compost windrows can be set up with virtually any residues from vegetable processing, straw, manure, and other organic materials when available in large enough quantities to make a row. Otherwise, with smaller quantities, it is preferable to make a single pile or to fill a compost bin.

Compost windrowing is an operation that can be practiced in any season with a variety of equipment. In normal farming situations, the piles are turned at least 1–2 times during the composting period. The amount and type of turning depends upon the size of the pile and the type of raw

Fig. 3.4. A farm compost windrow on the edge of a cropped field and with road access. Such piles may be continually added to with daily runs from a barn.

materials, as well as on factors such as temperature, moisture, and the planned use for the end product. Since many theories exist regarding the significance of compost turning it is advisable to make a practical determination of costs, process time, and end-use before making a decision.

TYPES OF WINDROWS

Compost heaps can be made in a number of different ways, from uniform windrows to stratified piles of multiple layers. When making it all at one time, care must be taken in sizing based on the overall porosity.

To protect a compost heap from drying-out or becoming too wet, it should be covered with straw or a suitable protective layer that allows the compost pile to breathe while shedding moisture or protecting it from evaporation. Several types of commercial synthetic fabrics are now available which provide this kind of protection.

Special windrows designed to save space and to improve aeration:

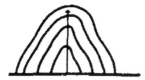

Fig. 3.5. Stratified layer piles
Various layers (4–8 inches thick) are spread one upon another at daily or longer intervals, moving from the center upwards. This form of pile heats very slowly but uniformly.

Fig. 3.6 – Consecutive layered piles
The principle is similar to the one above, except that the layers are spread like mini-heaps laterally on a daily or longer interval. Such a pile heats unevenly lengthwise depending on the age of the portions.

BUILDING AND TURNING WINDROWS

Compost building and turning is both art and science. While it is true that a well-built pile may not need much attention, it is also true that usually all piles need to be turned at some point or another. Contrary to popular opinion, a lot of turning has very little effect on the compost other than breaking down its texture. On the other hand, a pile that is never turned may develop a wet core lacking oxygen, leading to odors when turned or applied. This wet core does not heat up very much and

Fig. 3.7 – Manure spreader with side ejector, well adapted for building small to medium sized windrows.

therefore appears 'green' late in the process when the outer layers may be completely composted. By turning a pile, the 'dead core' effect is eliminated and the materials are homogenized.

In building a windrow, traditionally a pile is layered by ejecting materials with a manure spreader. Side delivery spreaders allow a pile to be dumped in place. With a rear-eject spreader some control of spread-width needs to be made.

It is possible to excessively turn a compost pile. Modern rotary-drum turning machines help reduce particle size, but if used too much will excessively and needlessly reduce the porosity and texture of the mix. If this happens, the ability of the compost pile to self-ventilate is restricted, often requiring more turning to compensate for it. This can become a downward spiral and is extremely undesired and not only drives up the cost of composting, but also hastens the losses of organic matter and especially nitrogen. The best approach is to turn as little as needed to maintain a loose, evenly heating mass.

For small quantities of compostable material, no special equipment is really required. Composting can be carried out on the soil surface, in a garden or on the edge of a field. In farm situations, it is recommended that compost piles be 'rotated' around the farm at different soil sites. By rotating composts, the soil is enriched below each pile but not to the extent that nitrate leaching or salinization becomes a serious problem.

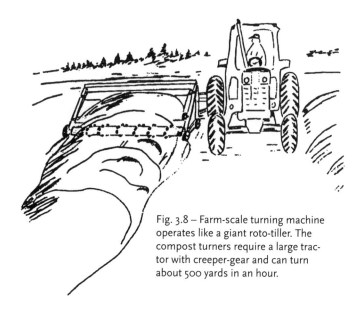

Fig. 3.8 – Farm-scale turning machine operates like a giant roto-tiller. The compost turners require a large tractor with creeper-gear and can turn about 500 yards in an hour.

Subsequent field cropping in these composted areas absorbs the excess nutrients and produces a bountiful harvest.

Two important compost research projects have been performed in which the type and extent of turning were compared to the extent of decomposition and stabilization of farm-based manure composts (Jobin and Brinton 1992; Brinton, et. al 1993). The graph on page 36 shows results from these trials evaluating the time it took compost to fall below 105 degrees F, an indication of stability, in comparison to the type and frequency of turning or the amount of straw-bulking material used during composting.

The test results from the trials illustrate that turning twice a week with a high-speed rotary turner improved stabilization by about twenty days over a total of one hundred days of composting; similarly, adding more straw to improve porosity increased the rate at which compost broke down regardless of turning. Not all compost may perform as well, but the research shows that with proper compost mix ratios, frequent turning may be of less significance to ultimate stability than believed.

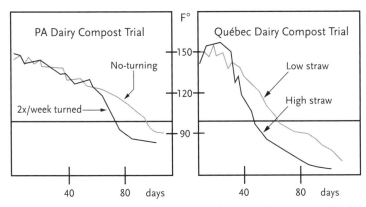

Fig. 3.9 – Relationship of turning frequency and straw addition on rate of stabilization of composted farm manure.

However, composts that were turned frequently appeared to be much more fine-grained and homogenous than did others.

Compost site design

Compost site design need not be complicated, but certainly depends on the nature and size of an operation. A simple site lay-out is comprised of a compost 'pad' consisting of a firm, well-drained base overlying dense semi-impermeable soil. For permanent sites, a suitable surface can be prepared with sieved, compacted gravel. The surface is inclined at a slight slope (1–2 percent) and fitted with a bermed collection area to retain any excess runoff. The retention pond can be used as a source of irrigation water, or the runoff water can be applied to the compost during dry spells. The swale and berm areas are planted to a conservation grass/legume mix.

If properly managed, a compost pad of this design will serve as a year-round surface with excellent load-bearing capacity of equipment. Such a site provides effective environmental control and meets most modern expectations of state regulatory agencies.

Equipment required in composting

SHREDDING

Pre-grinding is only required for coarse materials like brush and branches. Leaves may be optionally shredded but it is generally not necessary. A variety of garden shredders are available with limits on the quantity that can be usefully ground. For larger operations, bale busters and tub-grinders are available.

MIXING

Special mixing equipment is only necessary when dealing with unlike materials like sludge and sawdust or manure and straw. A manure spreader is particularly recommended for farmyard manure and fresh vegetable residues. Compost ingredients can be mixed by:
- Windrow turning machines (see below);
- Concrete mixers (batch mode);
- Feed mills (batch mode);
- Rotating mixer/pug mills (continuous feed).

WINDROWING

In most cases, ordinary farm machinery is suitable for building windrows, as for instance:
- Tractor fitted with hydraulic front-end loader;
- Manure spreader (with optional side ejector);
- Windrow machine.

For larger quantities (over 25 cubic yards), a front-end loader with 4-wheel drive may be needed. A variety of mechanical compost turners are also available for farm and larger scale use.

TURNING

Special turning machines are available, varying from farm-scale PTO-driven models to self-powered units for municipal or large farm operations. The turning machines are useful if large quantities of material have to be regularly processed and a faster time for composting is desired.

Fig. 3.11 – Dynamics of processing and tools used

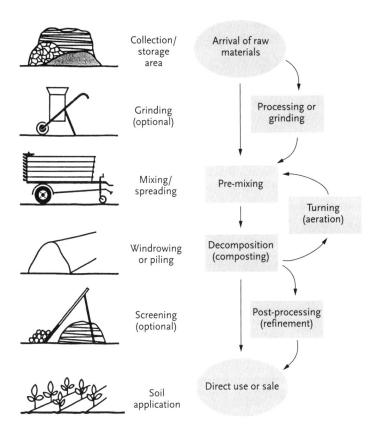

Sheet or Surface Composting

Sheet composting is an age-old practice, and contrasts with controlled composting in that the organic materials are allowed to break down on, or just under, the soil surface rather than being piled in heaps, bins or windrows. The term composting is, of course, a misnomer when applied to this practice. It is simply the natural degradation of wastes on the soil surface, a practice that has many drawbacks.

In order to conduct sheet composting, raw materials, such as harvest residues and green manures, are pre-chopped and mixed into the upper soil to a depth of about 2–6 inches (discing or rototilling). To allow the materials to pre-dry, they may be left on the soil surface to wilt after cutting.

Farmers often consider handling most crop residues in this manner to be superior to composting. Compost entails extra effort, and the wait period during composting may seem unnecessary. Surface or trash composting on the surface considerably reduces this wait.

Sheet-composting is not recommended for odorous, raw wastes like food scraps and certain manures. Furthermore, sheet composting is not possible without the right kind of mechanical equipment. The following machines are recommended both for grinding and soil incorporation:

- straw shredders;
- discs;
- rotovators or rototillers;
- rotary harrows with vertical shaft;
- spading machines.

Green manures or harvest residues should never be deeply incorporated with a plow, since they would putrefy rather than decompose, especially in heavy soils. Under adverse conditions, buried residues can be returned to the surface when plowing the following year. Too much carbonaceous material disced into the soil surface may also act to tie up nutrients for the following growing season.

After proper incorporation of a sheet-compost layer, the soil surface should appear rough or 'trashy'. The next or new planting should follow a waiting period to allow for decomposition. The length of waiting time depends on the soil temperature, the type of soil, the crop rotation, and the nature of the raw materials.

As a general guideline, the waiting time observed should be one-to-two weeks in summer and early fall, and two-to-three weeks in early spring and late autumn.

PRACTICAL EXAMPLE

In a market garden, cauliflower harvest residues (lower leaves and stalks) are left on the field. After harvesting, some form of nitrogen is applied (manure or concentrate) and lightly incorporated (about 4 inches deep) by means of a rotovator or disc. Soil preparation for the next crop may begin 10–20 days later (according to the degree of decomposition and the prevailing temperature).

The same method may be used for other types of vegetable residues (such as lettuce). The technique described above is especially for light and medium soils. The nitrogen applied, if needed at all, may take the form of any organic nitrogenous fertilizer, such as horn meal, liquid manure or compost.

Fig 4.0 – 'Sheet' composting on the soil surface – harvest residues before their incorporation by means of a rotovator.

Evaluating Compost Quality

Evaluating compost 'ripeness' or maturity

Composting is an intensive breakdown process, so it is natural to ask when is it done? Is it good for plants? As our earlier graph (Figure 1.2) indicates, the process goes through a warm or heating phase, a 'transformation' phase, and a final 'curing'. Somewhere in this process the compost may be considered 'ready to use'.

In order to appreciate the significance of compost maturity, consider the potential problems. An immature compost is one that has not completed the active breakdown phase and may have an inhibitory effect on plant growth, particularly with young plants and seedlings. This growth suppression effect may be caused by several factors. The ammonia may still be high, the material may demand oxygen and there may be the presence of phytotoxic compounds, e.g. organic acids formed from incomplete digestion of organic matter. The simple conclusion is: learn how to assess compost quality before a problem arises.

The most direct and reliable method of evaluating the quality of compost is to test it in the intended application, comparing its performance with that of a known and reliable standard in that specific application. The advantage of this approach is that all the aspects of quality that are important in the particular application will be evaluated together. The disadvantages are:

- Time and space might not be available for full-scale plant trials;
- The season might not be right for outdoor trials when the compost is ready to be evaluated;
- Compost quality might actually decline during the time it takes to conduct a full-scale trial.

These shortcomings can be overcome by using a small-scale, short-term simulation of the intended application, indoors if necessary. For example, compost intended for general landscaping could be blended

20–30 percent with sand and planted with several fast-growing annual seeds, perennial transplants, or sod plugs in six-inch pots in a greenhouse. A good 'control' for comparison would be an identical set-up using known high quality compost or a commercial soil-less potting medium in place of the test compost. Plant growth during a period of 2–3 weeks should give a good picture of the over-all quality of the compost compared to the known standard.

Maturity tests using plants

A quick and simple test of overall compost quality is a one-week assay of seedling growth of garden cress and a fast-growing grain such as wheat. The test compost is blended 1:3 with limed (1 percent by volume) peat moss. Two six-packs (one for cress and one for wheat) are filled with this blend (see Figure 5.0), and two are filled with a commercial soil-less planting medium for a comparison. Each cell is planted with ten seeds. After seven days, the growth is compared either by weighing or visually and the number of seeds germinated in each cell is recorded.

CRESS TEST

Germination and growth of cress in mature compost should be regular and completed two to three days after seeding. After seven days the seedlings should be green and showing white roots.

Fig 5.0 – The Cress Test – In the rear, a fully-cured compost blended with peat moss showing good germination and growth. In the front, an immature compost with poor germination and poor growth.

Fig. 5.1 – Wheat Growth Test
On the left is the control (immature compost); on the right is a example using well-aged compost

WHEAT TEST

Seeded in a mature compost peat mixture, wheat should germinate after three to four days and after seven days appear vigorous, green and healthy (no necrosis).

Fig 5.2 – Green Bean Test
Germination and growth of green beans after 10–12 days. On the left is the control (immature compost); on the right is the mature compost mixture.

GREEN BEAN TEST

Another less frequently tried test is to grow a legume (nitrogen fixer) in a 1:3 compost:peat/soil blend. Legumes are sensitive to excess ammonia and phytotoxic conditions and will not form nodules if the compost is immature. They are also very sensitive to certain chemical residues, displaying a dwarfing and puckering of leaf margins if herbicides are present at levels likely to effect crops.

The beans should germinate after five to seven days. Seedlings should stay erect and show a well-developed root system ten to fourteen days after seeding. Nodules should appear on roots after three weeks.

INTERPRETING THE SEEDLING ASSAYS

If a compost performs poorly compared to the control or standard in any of these trials, it may be difficult or impossible to determine what is wrong without measuring specific traits of the compost.

Generally speaking, poor germination compared to the control indicates phytotoxicity due to immaturity or high soluble salts content. Poor seedling growth can result from immaturity high salts content, or poor nutrient levels. A well-matured, rich, high-organic compost will outperform the control in these assays.

Some guidelines in understanding the qualities of 'ripeness' in relationship to the various tests that can be done to confirm it are as follows:

- Maturity associated with: mellow, gentle, stable, well-balanced, humified, microbially-diverse, low C:N, high soluble-N, neutral-to-acid pH;
- Stability: no longer rapidly decomposing, no nitrogen tie-up measured by decomposition rate, e.g. respiration, oxygen uptake, self-heating, oxidation-reduction potential;
- Soluble salts can 'burn' crops if too concentrated; measured by electrical conductivity;
- Available nutrients – depends on quality and ratios of starting ingredients, maturity, loss during composting.

Dewar Self-Heating Test

A test which bridges the gap between the technician and the practitioner is the Dewar Self-Heating Test, a procedure to standardize measuring residual heating ability of composts, found to correlate with maturity (see Figure 5.3).

The application of the Dewar flask to measure compost self-heating was first made in 1982 at the University of Stuttgart. This method for compost evaluation has slowly become one of the accepted standards in Europe (Becker 1995), and is being considered in several states in America (Brinton et. al. 1995). Presently, the Illinois EPA lists it as an acceptable test to determine if compost is 'finished' based on heating less than 20 degrees C above ambient in the flask.

Fig 5.3 – The compost self-heating test measures potential decomposition in a special two-liter Dewar flask equipped with a max-reading thermometer.

The objective behind the Dewar Self-Heating Test is to provide a simple, reproducible way to verify the potential for continued decomposition of composts, associated with some level of natural heating. The original concept is based on the assumption that organic materials have varying abilities to achieve certain levels of heating in proportion to their state of decomposition. Self-heating is important because it drives the compost process, and regardless of other traits, the presence of heat in compost is widely held to be a sign of immaturity. Thus, a compost that does not heat appreciably is considered to be stable or mature. The Dewar vessel is basically a super-insulated thermos that retains heat while allowing ample air to enter the sample for normal decomposition.

The principle of the method is to precisely record the highest temperature achieved with a maximum reading thermometer, after compost is placed into the vessel for several days. Interpretation of the results is based on division into five levels of 10°C increments of compost heating (see Table 5). For example, Class V refers to the least amount of heating, or up to 10°C, and Class I is the highest heating of 50°C or more over ambient. The test is slow and requires about two to nine days to complete.

Compost is prepared for the Dewar test by careful, representative sampling, cooling to room temperature and remoistening, if needed, prior to filling the two-liter vessel. Pre-screening compost to less than ¾" improves uniformity of results, but is not essential to performance.

Table 5.0 – Dewar Self-Heating Rating Scheme

Compost temperature rise above ambient inside the vessel	Class of stability	Description of compost stability
0 – 10° C	V	Very stable; well aged
10 – 20° C	IV	Moderately stable; curing
20 – 30 ° C	III	Active compost; material still decomposing
30 – 40 ° C	II	Immature, young, very active compost
40 – 50° C (or more)	I	Fresh, very raw compost

Table 5.0 shows how to rate the compost based on the temperature increments observed in the Dewar test. Producers may utilize the technique in the field provided a relatively stable ambient temperature of about 20–25°C can be maintained around the flask during the test period lasting about three to five days.

The self-heating test, like other tests, may suggest the compost is stable, when in fact it is not. The authors' experience is that if the compost has previously been too dry, too hot, or has a very low pH it may give a somewhat delayed heat reaction in the Dewar test. Thus, it is important to make sure the compost is in as 'normal' a condition as possible before conducting such a test. There are some useful general guidelines for using composts based on the Dewar scale, as seen in the following table.

Table 5.1 – Use of Compost Based on Dewar Ratings

Dewar stability class	Compost can at best be used for:
V	Potting mixes, seedling starters
IV	General purpose gardening, greenhouse cultivation
III	Grapes, fruit, apples
II	Field cultivation, e.g., corn, tomatoes, broccoli, greenhouse hotbeds
I	Compost raw feedstock; mushroom compost

Maturity tests using chemically reactive strips and reagents
It is possible to roughly determine the ammonia and nitrate values of compost and thus evaluate approximately its degree of maturity.

NITRATE TESTS

Experience shows that nitrate increases with the age of compost, all other factors being equal. It is possible to test the nitrate with simple soil-test kits or reactive strips and determine if this is the case. Nitrate values of well-matured compost will vary greatly but should generally be greater than about 200 ppm. Because nitrate is easily lost by leaching or as nitrogen gas under wet conditions, the tests must be read with caution.

AMMONIA TESTS

Ammonia is often present as a breakdown product from proteins and nitrogen compounds like urea present in fresh wastes and manure. The ammonia should disappear over time as the compost ages; in fact, most of it is converted into nitrate. No significant ammonia should be detectable in truly mature compost.

Table 5.2 – Maturity of a Compost Assessed by Nitrogen Levels

Compost stage	Nitrate (NO_3) level	Ammonia (NH_3) level
Degradation (phase I)	None or very low < 10 ppm	High > 1,000 to 10,000 ppm
Transformation (phase II)	Low to medium < 200 ppm	Medium to low – 100 to 1,000 ppm
Maturation (phase III)	Medium to high > 200 ppm	Low < 30 ppm

SOLVITA® COLORIMETRIC RESPIRATION PROCEDURE

The Solvita maturity test is a special Woods End procedure that shows grades of compost respiration in a system that changes color when exposed to compost in a jar. The test takes only four hours and requires a small two-ounce sample of the compost product. The material is placed into a special jar and exposed to a paddle that contains a chemically sensitive gel, which responds to the rate of biological activity in the sample.

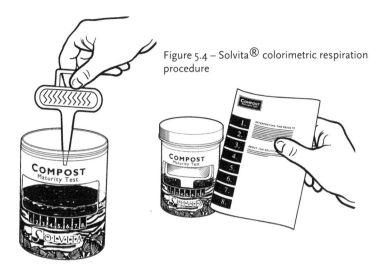

Figure 5.4 – Solvita® colorimetric respiration procedure

The color changes correlate with actual respiration rates (or oxygen uptake) as measured in the laboratory and also compare to the Dewar grades over the whole scale of color change.

Other laboratory analyses for quality and maturity

A variety of laboratory procedures exist whereby a compost sample may be evaluated for quality and maturity. Although the type of tests needed and the measurements offered by labs will vary, a few key traits need to be observed.

It is important to send a homogeneous sample from well-sampled and well-mixed material representative of the entire mass, similar to taking a composite soil sample. Usually a minimum final sample of about two pounds (one kilogram) is required at any specialized soil or environmental analysis laboratory.

Some of the types of tests that are important for assessing quality and maturity are noted in Table 5.3 on the following page.

Maturity or stability is not the only criteria for compost quality. While these terms encompass a wide range of qualities, there are still others that must be considered. These include but are not limited to nutrients, acidity/alkalinity, salt content, porosity, and the presence of

Table 5.3 – Laboratory Procedures for Evaluating Compost Maturity and Stability

Variable measured	Attributes determined	Stability indicated by:
Oxygen consumption	Uptake rate of oxygen needed to decompose remaining material	Low values
CO_2 respiration	Release rate of carbon dioxide from decomposing remaining material	Low values
Self-Heating	Heat released from remaining organic materials	Low values
Redox potential	Relationship of oxidizing to reducing chemistry in material	High values
$NO_3:NH_3$ Ratio	Conversion of nitrogen into more oxidized forms	High (>1) ratio
Humus Test Q4:6	Relative proportions of low to higher weight humus compounds	High ratio

weed seeds and plant pathogens. Several of these traits are customarily measured in a manner similar to testing soils.

Compost producers and users like gardeners and nurserymen often have their own rules of thumb and learn over time to correlate performance to various tests traits. The application of maturity and stability standards and their correlation to field performance is important to the successful use of composts in a variety of applications.

Compost Uses

Objectives

There are many excellent reasons to make compost, but like anything, the type of desired use must determine the proper application use.

FERTILIZING EFFECT

All composts possess some potential fertilization value, and with some it may be the factor that limits application. The nutrient content depends on the nature of the source materials. Compost also concentrates minerals.

SOIL IMPROVEMENT

Compost has a favorable influence on soil organisms and on soil texture, regardless of its nutrient value. For the ideal soil improvement, compost not too high in nutrients, but moderately mature, is generally best.

HUMUS EFFECT

Compost sustains and increases the content of soil humus. Its effect on the humus value depends on the type of compost used. Well-prepared bark compost is especially recommended for improving humus content.

PRACTICAL EXAMPLE 1

On a garden with light soil, compost is prepared from vegetable scraps mixed with manure and straw and stacked in piles. Within the same year the young compost is used as the winter soil cover.

PRACTICAL EXAMPLE 2

In a market garden, horse manure is mixed with mushroom compost and vegetable trimmings. The mixture is composted in windrows, which are turned twice a year. The mature compost is then used for soil improvement and as a fertilizer in greenhouses.

PRACTICAL EXAMPLE 3

Potting soil is prepared by allowing a mixture of dairy manure and straw to decompose for seven months in straw covered windrows which are turned several times. The mature compost is over-wintered in a cold-frame or root-cellar and used mixed with peat the following spring for a potting mix.

Compost for raising vegetables and ornamentals

The current use of peat and high organic soils for raising vegetables and ornamentals in market gardening and horticulture has become increasingly costly due to environmental concerns and transportation costs. For these reasons and others, home gardeners and market gardeners may be looking for alternatives to peat-based soils. Compost, properly prepared, may serve as a partial substitute.

Nurseries should use well-matured composts. A mature compost has no apparent phytotoxicity. A good nursery compost must also be low in salts. Composts and/or leaf composts have been steam-sterilized if necessary to eliminate weed seeds, but this practice is discouraged. A properly matured compost is naturally disease suppressive and sterilization is likely to reduce or eliminate this quality.

Compost for seed beds or potting soil

Well-matured composts that are also low in salts are best used for raising seedlings. A mature compost suitable for potting mixes has little apparent self-heating ability by the Dewar test. It should also test positive in nitrates. According to ingredients used for preparing the compost, it might be useful to balance with additional elements of low salt index (e.g., rock dusts, dolomite, and other trace elements).

In order to eliminate risk of nitrate or ammonia accumulation in potted plants, the growing media should not be pre-sterilized with steam. This mainly applies to lettuce and cabbage. Most seedlings are sensitive to salt levels, so the potting mix should be checked for salinity by the conductivity method prior to use.

Compost for cold frames
For use in cold frames, the compost should first be mixed well with the topsoil layer and then into the soil. The depth of burial depends on the level of product maturity. If the compost is very mature, it can be double-dug into the soil according to the French-intensive biodynamic principles. Early French growers double-dug immature compost into hotbeds so that the compost would warm the soil in the winter months.

Compost for greenhouses
Composts are used in greenhouses in order to maintain and increase the concentration of organic matter as well as nutrients to maintain the very high productivity expected of these soils. The nutrient value of the compost must therefore be taken into account. Of great significance is low salt content of the compost since heavy evaporation in greenhouses can create salinization problems from surface deposits. Plants vary widely in salt tolerance. The salinity is easily tested with a conductivity meter. Values over 3–4 mmhos/cm in the compost and 0.5–2.0 in the soil should be regarded with great caution.

In order to cover the soil in winter, while the greenhouses are empty, use compost in the early ripening phase, for instance semi-mature manure at a rate of two cubic yards per hundred square feet. Before planting and/or transplanting, mix it with the upper soil layer by means of a spading machine or rototiller.

During summer, apply either a compost in the early degradation phase or a manure compost to tomato or cucumber crops (maximum one cubic yard per 1000 square feet).

Compost for outdoor crops
Various kinds of material and methods of compost application can be used, and in some cases one can equally well practice sheet composting in combination with use of previously prepared composts.

During winter, care must be taken to keep soils covered either with green manure, cover crops, or compost. For compost distribution, a manure spreader with a spreading width of up to thirty feet is best suited, since it permits application without excessive driving over freshly plowed soil. If possible, avoid excessive transportation of raw residues. Let them decompose on the spot.

Compost for agricultural field crops

In principle, compost may be used for any kind of agricultural field crop. Compost is recommended for row crops at rates of up to twenty cubic yards (ten tons/acre), depending on nutrient content, and to pastures and hay/legume crops at up to seven tons/acre provided the material is very stable.

Compost can be used in combination with growing and turning under a green manure crop or liquid manure slurry. Manure compost is best applied during the autumn and spring, soil conditions permitting. For other types of crops the quantity to be applied is generally up to seven tons/acre.

PRACTICAL EXAMPLE 1

On a farm the manure is treated by windrowing with additions of mature compost. After a period of three to twelve months, depending on the purpose for which it is needed, the compost will be used as a fertilizer. Turning the compost piles will speed up the decomposition process, but this is not absolutely necessary.

PRACTICAL EXAMPLE 2

On a vineyard in Northern California compost is made from cow manure

Fig. 6.0 – Compost that is ready to spread may be stored in curing piles near the field where it will be used.

treated with biodynamic preparations. The compost heaps are protected from adverse weather conditions by an eight-inch layer of straw. When finished, after about two months, the compost is spread on the surface of the fields or disced lightly into the soil. This compost also serves as a basis for preparing compost teas for natural fungus protection according to a special procedure (Tränkner 1995).

Special uses for composts

Compost that is prepared from coarse ingredients, like bark with some grit or sandy content, is well suited for soil-erosion protection use. Under the circumstances, the compost should test low in salt so that it may be used at heavy rates, up to 100 tons/acre, surface spread.

Compost teas as natural fungicide

The potential disease suppressive characteristics of compost are the subject of increasing attention among researchers and growers. A recent emphasis is seen on anti-fungal properties of watery extracts of composts.

The use of watery compost extracts or compost teas differs from the reported use of solid compost used in container-media to suppress damping off and root-fungal diseases. Watery compost extracts are diluted

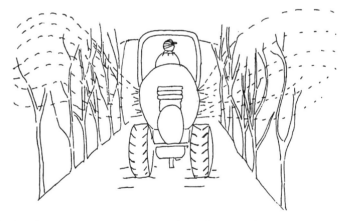

Fig. 6.1 – A compost tea extract is sprayed onto fruit trees during a period of fungus susceptibility.

'teas' prepared specially from finished composts and are applied directly to plant surfaces using a spraying apparatus. The extracts act directly in varying degrees to suppress both the germination and growth of plant pathogenic organisms (Tränkner 1993). The primary source of the effects observed with compost teas is of a microbiological nature. Sterilized or micron-filtered compost extracts exhibit little or no ability to impact pathogens. In other words, active biological control is apparently responsible for the positive effects observed for compost teas.

The most significant factors influencing the effectiveness of watery compost extracts are considered to be the age of the compost and the nature of the source ingredients. Composts of plant materials such as leaves, yard debris, and straw generally show reduced disease suppressiveness, with a loss of the ability after only three months of aging. On the other hand, composts containing manure (horse and dairy) show significant anti-fungal potentials up to nine to twelve months of age. In some cases, the antipathogenic effectiveness of the compost has been lengthened by drying the compost prior to storage, with the natural fungicidal properties elicited later after remoistening.

Use of compost for disease control should seen as a two-fold practice:
• Application to the soil surface to biologically combat soil-borne fungal spores;
• Use as a watery extract sprayed according to fungal infestation periods.

The actual composition and timing of events has to be worked out in accordance with accepted practice for the crops and disease in question.

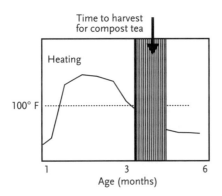

Fig. 6.2 – Selection of proper time to harvest compost for preparing watery-extract or tea for spraying plants.

Figure 6.2 depicts the relationship between the composting process and the time at which a suitable compost tea can be prepared. The actual point at which the extract should be made varies somewhat with the type and nature of the composting process.

The graph shows that at a certain point in a moderately advanced composting process, the material is ideally suited for preparing the compost tea. In general, the Dewar stability should be stage IV or higher (see Table 5.0), which translates into a compost pile that is heating to less than 100 degrees F. The actual age of the compost at this point may vary from as little as sixty days up to six to seven months. Directions for preparing the teas include steeping for five to eight days in water, then filtering and spraying without further dilution.

The successful use of compost teas depends on a proper appreciation of the relationship of the fungal agent to weather patterns and plant susceptibility. In some cases, the teas will have to be sprayed often after moist weather to prevent mildew outbreaks. Thus, it will be necessary to prepare compost batches to provide a continuous source for teas over the entire growing season. A final point is that certain wetting agents or surfactants are harmful to bacteria; therefore selection must be carefully made.

Compost Mulching

Unlike the sheet composting technique in which the materials are slightly incorporated into the soil surface, mulching consists of leaving the materials on the soil surface. Mulching composts are therefore considered to be those low in nutrients and coarse and woody in texture.

The advantages of mulching
- Regulation of soil temperature;
- Regulation of soil humidity;
- Decrease of erosion;
- Improvement of soil biological activity;
- Control or inhibition of weed germination;
- Improvement of soil structure.

Mulching fruit trees
Mulching with fresh organic residues as well as composts is recommended in nurseries and in orchards in accordance with regional practice. Often, large quantities of compost are used in nurseries, whereas mulching

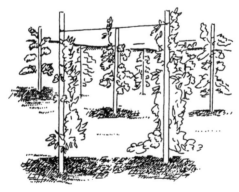

Fig. 7.0 – Orchard with a soil cover of bark composted under the trees.

is preferred in orchards. For this, one can use equally well materials in which the decomposition has already started or fresh materials, such as chopped pruning residues, grass clippings, and leaves.

PRACTICAL EXAMPLE 1

When planting trees in the orchard, the rows are covered with a 1–2 inch thick layer of compost. This warms the soil and provides nutrients in a slow release manner. Additional applications with compost are made every two to three years.

PRACTICAL EXAMPLE 2

A two-inch thick layer of chopped bark (maximum particle size of one inch) is applied along the tree rows. An herbicide effect can be observed for up to two years. It should be noted that bark should be applied only on a clean soil that is free of weeds.

Partially composted bark and municipal compost are particularly suitable for mulching. Mature compost does not generally have any herbicidal effect on weeds.

In viticulture

For vine growing, one can use various mulch materials, as well as mature compost. This ensures a fast warm-up of the soil in the spring and

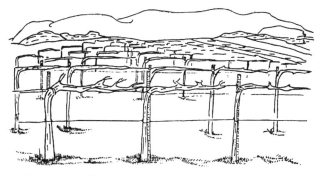

Fig. 7.1 – Vineyard use of compost – the compost is applied both in the rows to nourish the vines and between the rows to provide fertility to cover crops and to reduce fungal spore production on ground-litter

a reduction of erosion on slopes. Compost spread between the rows provides an inoculum to aid control of soil-borne spores that lead to fungal attack on leaf surfaces (Brinton 1997).

PRACTICAL EXAMPLE

In a vineyard, planted on a slope in the region of Karlsruhe in Germany, bark has been used as a mulch. Besides a decrease of erosion and improvement of the vegetation, an increase of the sugar content has been noted in the grape juice.

In shrub nurseries

Nurseries go through large amounts of materials rich in organic matter. Materials generally include peat, peat soil, wood products and composts.

Compost is recommended in various dilutions for:
- Propagation of young plants;
- Soil cover and soil amendments;
- Container media in mixes with sand and peat.

Soil cover (mulching) with fresh, decomposing bark, or with fragments of ornamental bark is used to:
- Improve soil structure;
- Provide weed control;
- Contribute to ornamental value (aesthetic effect).

PRACTICAL EXAMPLE 1

A mixture of peat, horse manure, and cattle manure is composted for six to twelve months for use outdoors on intensive growing plots.

PRACTICAL EXAMPLE 2

A mixture of one third decomposing bark, one-third peat, and one-third rock wool is used as a potting soil.

PRACTICAL EXAMPLE 3

Unsold or substandard products, as well as easily-degradable wastes, are composted for one year and then scattered over the fields to form a two to four inch soil-building layer every two to four years or so.

Composting for Kitchens and Family Gardens

Basic principles

Kitchen, household, and institutional composting begins with clean separation. In the unfortunate practice of modern western society, most household wastes are mixed and disposed of in landfills. Modern kitchens are not designed for separation of compostable from non-compostable materials, a practice that requires extra space and bins.

A variety of means exist to start the process, but a chief factor is convenience and clean separation. An idea that originated in Europe is the use of special aerated pails or compostable paper sacks to collect food scraps. For large kitchens or institutions, biodegradable liners can be used for regular trash containers. Plastic is to be avoided!

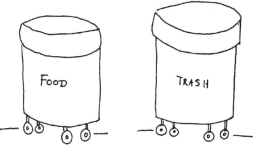

Fig. 8.0 – Separate bins are needed to encourage separation of food from non-compostable materials.

Situate the compost area in a location that is convenient but not a nuisance to the users. It is not unusual to observe compost bins in close proximity to a kitchen rear door, or a short walk across a schoolyard, possibly next to a maintenance shed where tools are kept.

For home composters, perhaps the best location is near the center of the garden. However, in order to get family members to participate a

Fig 8.1 – The FoodCycler™ is a compostable bag made of recycled kraft paper lined with a leak-proof cellulose liner used to collect food scraps for home or municipal composting.

convenient location for the compost pile is mandatory. It is far easier later to carry the finished compost to where it is needed than to make a bi-weekly trek with the food scraps (in all weather!) to the bin. In any event, with pre-formed bins or loose piles, composting may be moved around depending on need and circumstances. It is certainly necessary to allow a sufficiently large area for the compost heap, and it is unrealistic to expect it to have the tidy look of a groomed lawn, particularly if one is serious about recycling branches and shrubs, while accommodating operations like grinding and sieving.

As has already been pointed out, in order to ensure a proper composting process, care must be taken to mix together different materials (for instance: rich in nitrogen/poor in nitrogen; fresh plants/wilted plants; coarse structure/fine structure; dry material/moist materials, and so on). When working with large amounts of kitchen scraps for instance, a sufficiently large amount of stockpiled dry bulking matter, like leaves or hay, is essential. The school compost programs designed by Woods End Laboratory depend on baled hay, leaves, and limestone as significant ingredients along with the cafeteria scraps (MWMA 1994).

It cannot be emphasized enough how composting will suffer if too much of the same materials, like food scraps, are piled together. What happens? The compost may become too wet, begin to attract flies, and will certainly develop the typical garbage odor associated with organic-acid type fermentation.

The more balanced and diverse the mixture of raw materials added

to food scraps is, the better the composting process will be. At this point, the average moisture content of the compost pile is of prime importance. In order to check the correct moisture content of the pile, the following practical test is recommended:
- Take a homogeneous sample of compost and squeeze it strongly. At most only a few drops of water should appear between your fingers.
- If the water runs out, the mixture is too moist, and some dry material, such as straw or old hay, should be added.
- Alternatively, if the sample crumbles when opening your fist, the mixture is too dry and must be moistened (a nettle slurry is particularly recommended for this).

In order to protect the compost heap from the weather, if it is not inside a shed, cover it with a layer of straw, a canopy, an air-permeable fabric, or at worst, plastic sheeting. In the latter case, inserting wooden spacers between the pile and the plastic will provide for ventilation between the heap and the plastic layer. The covering is simply thrown aside for pile mixing.

Fig. 8.2 – A composting shed situated near a school cafeteria. The building contains four bins and needed tools and materials for mixing with food scraps.

Special Composts

There are many ways that composting can and should be tailored to special needs. On the other hand, it may be very useful to customize a compost process based on residues or plant remains that are available in large amounts. In this case, there are no ideal recipes, but the following list includes some of the world's gardeners' favorite special composts.

Nettle compost
According to experienced and professional gardeners, stinging nettles make one of the best composts available if you like to cut them. Two thirds of chopped fresh nettles are mixed with one third earth or a manure compost. The beneficial action of nettles in such a compost is greatest when the nettles are harvested before flowering.
- For plants having high nutritional requirements and as a compost activator.

Soil compost
Soil compost consists of a large percentage of earth, and is therefore low in nutritive elements. It can be made of sod turfs, pond dredgings, or mud from ditch cleaning. Layers of sods are stacked roots upwards, separated by layers of earth compost (varying in thickness up to several inches) or nettle compost powdered with some calcified seaweed.

When preparing soil composts, pond dredgings, as well as mud from ditch cleanings, are first spread out to dry and then mixed with coarse structured materials (chopped branches, straw etc.) before being added to the heap.
- Soil composts based on turf are mainly used in horticulture, whereas composts from pond dredgings and ditch cleaning muds are used in market gardening. The pH is usually about neutral (around pH 7). Also recommended to replace peat or peat compost.

Leaf compost or leaf mould

The composition of the various leaf species varies greatly. Beech leaves are rich in calcium, whereas oak leaves will produce a compost with a lower pH. Leaves of birch trees are particularly useful.

In order to prevent the leaves forming a compact mass during composting, leading to a poor circulation of air, it is advisable to first chop them or pass them through a lawn mower or shredder and then to mix them with soil or compost. They should be allowed to decompose for at least one year.

- For potted plants and nursery plots. Oak tree compost is recommended for ericaceous plants, such as azalea or rhododendron.

Manure compost

A well-made, mature farmyard manure compost is about the most ideal garden fertilizer. When mixed with other composts as well as with soil and peat, dairy manure compost is to be highly recommended for starting plant seedlings. Each type of manure (e.g., cattle, pig, poultry) has its own composting characteristics. Cattle manure contains the most ideal balance of nutrients and can usually be composted by itself. Pig manure needs to be mixed with lots of straw and other bulking agents. Poultry manure needs to be blended with large amounts of carbonaceous materials such as straw, sawdust or leaves.

In European traditions, cattle manure is often first mixed with some soil or an activating compost, plus rock powders or other additives (see section on additives). In the likelihood the manure does not contain sufficient bulking material (hence too moist), it is advisable to add some straw or leaves.

To prevent excessive heat when composting manures with a high temperature potential (e.g. horse manure), the heap should not exceed a height of about four feet. Keeping piles reasonably small means that the generated heat will be able to dissipate more easily. Very large piles need to be specially turned or aerated.

- As a garden soil amendment and fertilizer for heavy feeding crops; as a seedling starter (after mixing with garden soil at a 1:1 ratio).

Bark compost

Bark and the by-products of the wood and lumber industry are local materials, which when correctly composted can replace peat (imported material). Use only materials that were not previously treated with insecticides or preservatives.

When mixed with materials rich in nitrogen, bark may be composted like straw. Commercial bark composts are widely available and have been shown to have disease suppressive potentials. Bark blended with poultry manure makes very good compost. When mixed with a rich clay soil, this kind of compost will produce an excellent potting mix for plants and flowers. When compared with peat, bark compost has the advantage of a more varied structure (particles of different sizes) and a better pathogen control effect.

'Activating' or inoculum composts

Activating compost is a somewhat vague term, which refers to any compost that is specially prepared to be re-used as a kind of inoculum in composting. Occasionally, minerals and herbal plants or other treatments are also applied. Such an activating compost is used to initiate or stimulate the composting process in fresh compost heaps. An example of this is the compost called 'barrel compost' used in biodynamic circles, or curing compost that is recycled back into the new piles.

Screened residues from matured composts can successfully be used as activators, but little or no evidence exists which proves their usefulness and efficacy. It is thought that such composts contain all the microorganisms necessary for a proper composting process. A good idea is to add such activators to the compost after the high heat phase to help re-establish the desired microflora.

Straw and hay composts

As with leaves, straw and to a lesser extent spoiled hay, do not decompose very readily, but will eventually make a fine compost. Oat straw is considered to be an exception, with a much faster rate of deterioration. In most cases, it is helpful but not essential to chop the straw before composting. Another technique is to take the straw bales and immerse them in water or manure slurry for several days before composting, or simply expose the chopped straw to rainfall in thin layers. Well-moistened straw can be mixed in the proportion two-thirds straw with one-third soil or an activating compost and it will commence composting.

In some European traditions, calcified seaweed and rock phosphate are also added to young hay composts. The traditional approach when piling up the heap (and later during composting) is to sprinkle a mixture of diatoms and rock minerals along with a liquid manure extract prepared from poultry manure and water, up to a level of about one-half pound of dry poultry dung per two pounds of straw.

- This type of supplemented straw compost is especially recommended for small fruits (berries) and shrubs. When it contains a little poultry dung and is well stabilized, it can also be recommended for onions and carrots.

Grass-clipping/spoiled hay composts

It is not always necessary to mix grass with carbonaceous material before composting. Swiss landscapers windrow grass after it wilts by heaping it into small piles less than two feet high. These little piles may get extremely warm, and start a very rapid rotting process, culminating in a dark, friable compost. They can be easily flipped with a hayfork or even a tedder machine. The small size prevents the odor associated with large grass piles.

Grass compost prepared in the above method and allowed to compost for about six weeks, can be heaped into larger piles and stored in a cooler, moist condition, such as a cold-frame. The experience of professional gardeners suggests that these pure grass composts make premium seedling starter mixes when combined with about 50–60 percent coir, peat, or soil for the next season.

Unchipped wood composts

Grinding or chipping branches and pruning is a time-consuming, energy-consumptive and noisy task. Swiss composters have demonstrated that the extra step of shredding prior to composting may not be needed, if time and space are on your side.

In one Swiss method designed by landscapers, there is no grinding whatsoever, and branches and limbs are piled or heaped into windrows and then trampled, either manually underfoot, or with a small tractor that drives part-way onto the piles. Natural weathering forces, including dry-rot and enzymatic cell-wall destruction, will significantly increase the brittleness of the wood over time. When the pile is re-trampled at a later date, it will collapse much more significantly. Meanwhile, the small pieces of bark and wood that slough off during this physical comminution process will become colonized by macro-fauna, fungi and bacteria, and eventually a true composting process under the coarser wood. By lifting the piles somewhat, a fine woody humus-like mass of compost accumulates over a one-to-two year period. Add this compost to container media or around shrubs for a great effect.

Glossary of Technical Terms for Compost and Soil

Actinomycetes
   Family of micro-organisms often confused with fungi belonging to a group intermediary between bacteria and fungi, but considered to be true bacteria.

Aerobic
   Of or related to microorganisms capable of using oxygen or other electron acceptors to complete the oxidation of organic matter.

Anaerobic
   Microorganisms capable of metabolism in the absence of oxygen and associated with fermentation processes.

Ash
   Mineral component residue remaining after combustion or decomposition of organic matter, often present in large amounts in aged composts.

Bentonite
   Special type of clay (also see Montmorillonite).

Biodynamic Method
   An agricultural method stressing whole-system farming and composting, founded in 1924 by Dr. Rudolf Steiner, an Austrian philosopher-scientist.

Biodynamic® Compost Preparations
   Naturally formulated products used in the Biodynamic cultural method with the purpose of influencing the composting process. The herbal plants are specially prepared prior to use and are referred to as preparations:
- 502: Yarrow (or Milfoil)
- 503: Chamomile

- 504: Nettle
- 505: Oak bark
- 506: Dandelion
- 507: Valerian

C

Chemical symbol for carbon, comprising 54–58 percent of organic residues.

Carbohydrates

Compounds including sugars, starches and cellulose. Sugars are generally assimilated by bacteria.

Cellulose

Carbon component of plants, not easily digested by microorganisms.

C/N ratio

Ratio representing the quantity of carbon (C) in relation to the quantity of nitrogen (N). This factor is often used in determining the composting potential of a material.

CO_2

Carbon dioxide, the principal by-product of oxidative respiration by microbes of organic matter.

Decomposition

Conversion of organic matter as a result of microbial and/or enzymatic attack; resulting partly in humus and CO_2.

Degradation

See *Decomposition*.

Dewar

A physical apparatus used to determine self-heating potential of compost and used in Europe for identifying mature composts.

Dolomite

Geological formation rich in calcium and magnesium, used as a fertilizer in agriculture.

Disease suppression

The quality of condition of a compost inducing control of pathogenic organisms, resulting either from antibiosis, competition or aggression against an undesired microbe.

Edaphon
All the organisms (plants, animals and micro-organisms) living in or on the soil surface.

Enzymes
A unique class of proteins that catalyze reactions; present in large amounts as a result of compost microbial activity.

Facultative
An organism that can switch between aerobic and anaerobic metabolism in response to environmental conditions.

Fermentation
The opposite of aerobic respiration; partial or incomplete oxidation by microbes of organic compounds in absence of oxygen; often associated with release of odor.

Gypsum
A natural compound, $CaSO_4$, used to control excessive loss of ammonia from manure and composts.

Heavy Feeders
Plants (like corn, tomatoes) with high requirements of available nutrients such as nitrogen and potassium.

Heavy Metals
Elements such as Cu, Cr, Hg, Pb and Se generally considered to be undesired residues in wastes.

Inoculum
Any substance or material considered to be useful to supply needed microorganisms to a biological process.

Lignin
A substance involved in the structure of wood, responsible for its rigidity and which decomposes very slowly if at all in soil and compost.

Lipids
A generic term for all fats, oils and related fatty compounds often present in large amounts on food and meat by-products.

Macro-nutrients
Nutritive elements (like N, P, K) needed in large quantities to ensure normal plant development.

Maturity
A condition of compost in which decomposition of organic material is extremely slow, self-heating is very low and soilcrop effects are favorable; see stability.

Mesophilic
The quality or state of microorganisms growing at normal temperature ranges of about 5°C up to 48°C, typically encountered in moderately active and finished composts.

Micro-nutrients
Nutritive or trace elements (like Fe, Cu, Zn, Mn) needed in small quantities for healthy plant development.

Microbe
Microorganism belonging to a diverse class of single-celled microscopic organisms which include bacteria, fungi, yeasts, algae and protozoa ranging in size from about 0.5 to 5 microns.

Montmorillonite
Component of bentonite consisting of swelling clay particles with a high surface exchange capacity, sometimes added to compost and soils.

Mulching
Method of covering the soil surface with raw or composted organic residues like bark, wood chips and peat.

MSW
Municipal Solid Waste: the mixed residue or trash component of town waste including paper, cardboard, garbage and recyclables, which must be sorted prior to composting.

Necrosis
Dieback symptoms in plants, such as yellow spotting; often caused by phytotoxic composts.

Phytotoxicity
The condition or state of suppressing plant growth; usually associated with volatile organic acids in composts; also related to high levels of ammonia, salts and heavy metals.

Respiration
The opposite of fermentation; aerobic metabolism yielding free energy

(heat) and resulting in formation of carbon dioxide and water after break-down of organic matter.

Rhizoctonia
A genera of fungi which grows on organic matter and which may cause damping-off in sensitive seedlings; often present in unripe composts but suppressed by mature composts.

Trace elements
See micro-nutrients; also heavy metals.

Soil exhaustion
An impoverished state of the soil in which crop plants cannot develop properly; associated with build-up of pathogens, phytotoxic compounds (often resulting from unsuitable cropping patterns) and the loss of soil structure and nutrients.

Stability
A condition in compost associated with very low respiration rate; often equated with maturity.

Thermophilic
The quality or state of microorganisms being resistant to high heat, generally considered to be temperatures greater than 48°C, often encountered in active composts.

Volatile Acids
Organic acids formed during breakdown of fats and from anaerobic fermentation of wastes; associated with odor and phytotoxicity.

Waste/Residues
Organic materials or organic by-products associated with trash that may be suitable for making composts.

Compost and Related Bibliography

Dynamics of the Compost Process

Bezdicek, D. 2001. Persistent herbicides in compost. *BioCycle* 42:25–30

BioCycle editors. 1991. *The art and science of composting.* Emmaus, PA: JG Press, Inc. (Compendium of reprint articles.)

Bockemühl, J. 1988. On the life of the compost heap. In *In partnership with nature.* Wyoming, RI: Biodynamic Literature.

Brinton W. 1979. Fertilizer effects on humus quality. *Compost Science Land. Util.* September/October.

Brinton W. and E. Evans. 2002. Herbicides in compost: potential effects on plants. *Composting News* (April). <www.recycle.cc>.

Christopher, T. and M. Asher. 1994. *Compost this book.* San Francisco: Sierra Club Books.

Goldstein, W. 1982. A contribution of the development of tests for the biodynamic preparations. *BIODYNAMICS* 142:3–20.

Hoitnik, H and H. Keener, editors. 1993. *Science and engineering of composting.* Wooster, Ohio: Ohio Agricultural Research and Development Center, Ohio State University.

Jobin, P. and W. Brinton. 1993. *Effects of compost handling on quality parameters.* CDAQ, St. Elizabeth, Quebec. Report to Quebec Ministry of Environment.

Linder, M.L. 1984. *Nutritional biochemistry and nutrition.* London: Elsevier Press.

Mustin, M. 1987. *Le compost, gestation de la matiére organique.* Paris: Editions François DUBUSC.

Parnes, P. Robert. 1990. *Fertile soil: a growers guide to organic and inorganic fertilizers.* Davis, CA: AgAccess.

Rhode G. and F. Schneider. 1956. *Stalldünger und Bodenfruchtbarkeit.* [*Farmyard manure and soil fertility.*] Berlin: Deutscher Bauernverlag. 141.

Sauerlandt W. 195). *Stallmistkompost.* [*Farmyard compost*] Landwirtschaftangew. Wissenschaft, Heft 57.

Vogtmann, H. and J. Besson. 1978. European composting methods: treatment and use of farmyard manure and slurry. *Compost Sci./Land. Util. (Biocycle)* 18:15–19.

Quality and Maturity Standards

Becker, G. and K'ter. 1995. A standard measurement for compost maturity. *Biological waste management: a wasted chance?* (Conference proceedings, April 4–6, 1995). Bochum: University of Essen Press.

Brinton, W., E. Evans, M. Droffner and R. Brinton. 1995. Standardized test for evaluation of compost self-heating. *Biocycle.* (November) 64–69.

Iannotti, D. A., T. Pang, B. L. Toth, D. L. Elwell, H. M. Keener and H. A. J. Hoitink. 1993. A quantitative respirometric method for monitoring compost stability. *Compost science & utilization* 1(3):52–65.

Mathur, S. P. et al 1993. Determination of compost biomaturity: literature review. *Biological agriculture and horticulture* 10:65–85.

Mecklenberg, R. 1993. Compost cues: how to evaluate and use urban waste compost for plant production. *American Nurseryman* (February 13).

Spohn, F. 1978. Determination of compost maturity. *Compost science/ land utilization.* (May/June) 26–28.

Food, Farm, and Community Collection and Wastes

Beyea, J and M. Conditt. 1993. *Wet bag composting demonstration project, Greenwich and Fairfield, CT.* (Miscellaneous report; unnumbered). National Audubon Society.

Brinton W. and M. Droffner. 1994. Microbial approaches to the characterization of composting process. *Compost Sci. Util.* 2. Emmaus, PA

Brinton, W., E. Evans, and J. Collinson. 1992. *On farm composting: guidelines for use of dairy and poultry manures in composting formulations.* Report to USDA Soil Conservation Service, Chester, PA. (Available from Woods End Laboratory).

Brinton, W and W. Seekins. 1990. *Composting potato culls and potato processing wastes: a feasibility study.* Mt. Vernon, ME: Woods End Research Laboratory.

———. 1988. *Composting fish by-products: a feasibility study.* Waldoboro, ME: Time and Tide RC&D – Mid-Coast Compost Consortium.

Cato, J., editor. 1992. *Composting and using by-products from blue crab and calico scallop processing plants in Florida.* Gainesville, FL: Florida Sea Grant College Program.

MWMA. 1994. *The Cape Cod Hill School compost program: manual and operations.* Mt Vernon, ME: Woods End Research.

Rynk, R., editor. 1992. *On-farm composting handbook.* Ithaca, NY: Northeast Regional Agricultural Engineering Service, Cooperative Extension.

Woods End Resarch Laboratory. 1991. *Newsprint bedding: compostability, health risks and biodegradation.* (PAH). Mt. Vernon, ME: Woods End Research Laboratory.

Value and Use of Compost Products

Brinton, W. 1997. Compost for control of grape powdery mildew (*Uncinula necator*). *BIODYNAMICS* (Spring).

Brinton, W. and D. Tresmer. 1986. Organic growing media: use of compost in potting mixes. *IFOAM Conference Proceedings.* Santa Cruz: University of California. (Available from Woods End Research Laboratory, Mt. Vernon, ME.)

Chaboussou F. 1978. La résistance de la plante vis-à-vis de ses parasites. [Plant resistance in relation to parasites]. *Proceedings 1st International Research Conference*: 56–59. (IFOAM, Wirz-Verlag, Aarau.)

Cook R. J. 1976. The oxygen-ethylene cycle and the value of compost. *Compost Sci./Land Util.* 17: 23–24.

Eichenburger M., P. Ott, P. Schudel, and H. Vogtmann. 1981. Über den einfluss von kompost und NPK-düngung auf ertrag und nitratgehalt von spinat, schnittmangold, kopfünd n?sslisalat. [About the influence of compost and NPK-fertilization on yield and nitrate content in spinach, leaf-beet, lettuce and cornsalad.] *Mitteilung. Lebensmitt. Hyg.* 72: 31–43.

Hoitink, H. A. J. 1980. Composted bark: a lightweight medium with fungicidal properties. *Plant disease* 64: 561–565.

Lairon D. 1984. *Effects comparés des fertilizations organiques et minérale sur plusieurs paramêtres nutrionnels de productions maraichères.* [Comparative effects of organic and chemical fertilizers based on nutritional parameters.] INSERM, Université d'Aix-Marseilles II.

Maynard, A. 1993. Leaching of nitrate from compost amended soils. *Compost Sci. Util.* 1 (2): 65–73.

McLeod, E. 1982. *Feed the soil.* Graton, CA: Organic Agricultural Research Institute.

Petterson, B. D. 1984. *Organic, biodynamic and conventional cropping systems: a long term comparison.* Mt. Vernon, ME: Woods End Agricultural Institute.

Reganold, J. 1993. Soil quality and financial performance of biodynamic and conventional farms in New Zealand. *Science* 260:344–349.

Schaerffenberg B. 1968. Der Einfluss der Edelkompostdüngung auf das Auftreten des Kartoffelkäfers (Leptinotarsa decemlineata). [The influence of composted manure on Colorado beetle populations]. *Z. angew. Entomol.* 62: 90–97.

Schmid, O. and R. Klay. 1989. *Green manuring: principles and practice of natural soil improvement.* Mt. Vernon, ME: Woods End Research Laboratory.

Schuphan, W. 1972. Effects of application of inorganic and organic manures on the market quality and biological value of agricultural products. *Qual. plant. mater. vege.* XXI: 381–398.

Tränkner, A. 1995. *Compost and plant health.* Kimberton, PA: Biodynamic Agricultural Association Press (Summer, 1996 publication date).

Vogtmann, H, K Fricke, B Kehres, and T. Turk. 1989. *Bioabfall Kompostierung.* Witzenhausen: Hessen Ministry for Environment.

Vogtmann, H., K. Matthies, B. Kehres and A. Meier-Ploeger. 1993. Enhanced food quality: effects of compost on the quality of plant foods. *Compost sci. util.* 1 (1): 82–101.

Weltzien, H.C. 1991. Biocontrol of foliar fungal diseases with compost extracts. In J.H. Andrews and S. Hirano (ed.), *Microbial ecology of leaves.* 430–450. Brock Springer Series in Contemporary Bioscience, BSBN 038797579-9.

Useful Addresses – Technical Assistance

German-Swiss Edition Contributors
- Hirschheydt, Arnold von, Dipl. Ing. Agr Kornstrasse 1, CH-8603 Schwerzenbach, Switzerland
- Ott, Pierre, Ing. Agr ESA, Directeur, Institut technique de l'agriculture biologique, F-67600 Selestat, France
- Pfirter, Alex, Ing. Agr., Adviser Coop•rative Migros Aargau/Solothurn, P.O. Box 241, CH-5034 Suhr. Switzerland
- Vogtmann Hardy, Prof. Dr. Ing. Agr.ETHZ, Gesamthochschule Kassel, Nordbahnhofstrasse 1a, D-3430 Witzenhausen, BRD.

English Edition Translation/Editors
- Brinton, William F., Woods End Research Laboratory, Inc. Old Rome Road, Mt. Vernon, ME 04352
- Brinton, Richard B., 37 Chandos Road, Stroud, GL5 3PU. U.K.

Compost Information Sources
- Biodynamic Farming and Gardening Association. P.O. Box 29135, San Francisco, CA 94129-0135
- The Composting Council, 114 South Pitt Street, Alexandria, VA 22314
- Rodale Institute Research Center, 611 Siegfriedale Road, Kutztown, PA 19530

Compost Newsletters
- *Composting News*, 8383 Mentor Avenue, Suite 102 Mentor, OH
- *Compost Matters*, Woods End Institute, PO Box 297, Mt. Vernon, ME 04352
- *Biocycle*, JG Press, 419 State Street, Emmaus, PA 18049

Compost Gardening Supplies
- Gardener's Supply Co., 128 Intervale Rd., Burlington, VT 05401
- Johnny's Selected Seeds, Foss Hill Road, Albion, ME 04910
- Smith & Hawken, 117 East Strawberry Drive, Mill Valley, CA 94941

Production Notes

Cover photograph by William F. Brinton

Text illustrations by Margaret B. Collinson

Layout and production by Bruce Bumbarger

Set in Adobe Garamond and Berthold's Scala Sans

Printed by Thomson-Shore, Inc.